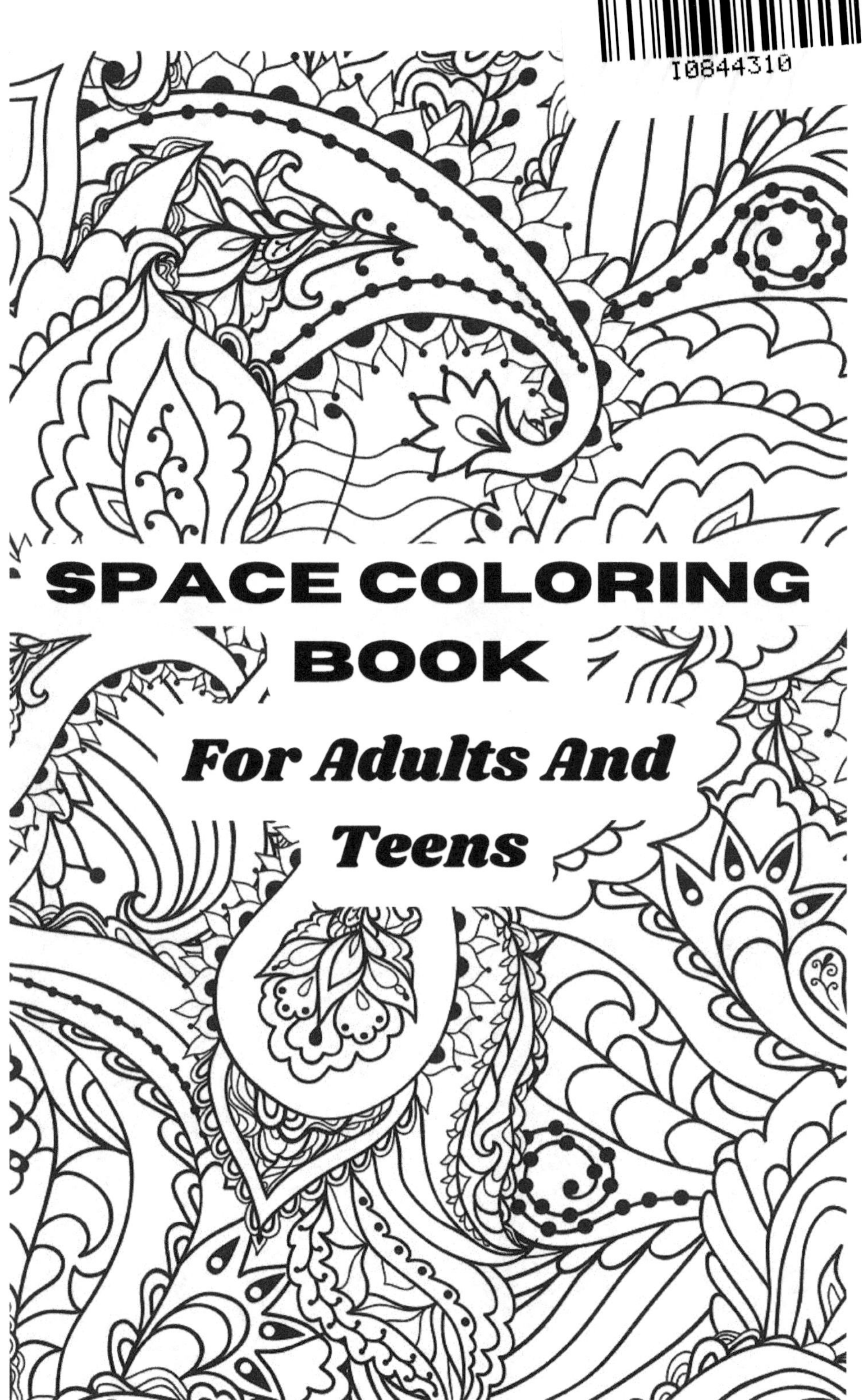
SPACE COLORING BOOK
For Adults And Teens
I0844310

This Book Belongs To

How To Use This Book

Greetings from across the universe! This coloring book about space is your passport to the wonders and beauty of the cosmos. Here are some tips for maximizing your cosmic trip, whether you're an adult or a teenager:

Start by leafing over the pages to become inspired. Allow your curiosity and imagination to be piqued by the breathtaking cosmic drawings.

Select Your Equipment: Take out your preferred coloring materials, such as watercolors, markers, or colored pencils. Accept your creative side.

Look around and unwind: Pick a page that appeals to you. Explore the minute minutiae of cosmic events, galaxies, and planets. Find inner calm and let go of tension as you paint.

Customize Your Space: Give each illustration a distinct personality by using color selections. You want an exotic planet with lavender skies? Try it out! This is your canvas in space.

Learn as You Go: Every page offers an opportunity to learn amazing space-related knowledge. While you go, read the subtitles and learn a little bit about space science.

Share Your Creations: Your completed works of art, whether you are an adult or a teenager, should be seen to others. Share them on social media, give them to friends, or just have fun with them.

Recall that space coloring has no set rules. One step at a time, travel the universe with this book as your pass. I hope you have an incredible adventure among the stars and happy coloring!

Taste Your Color Here

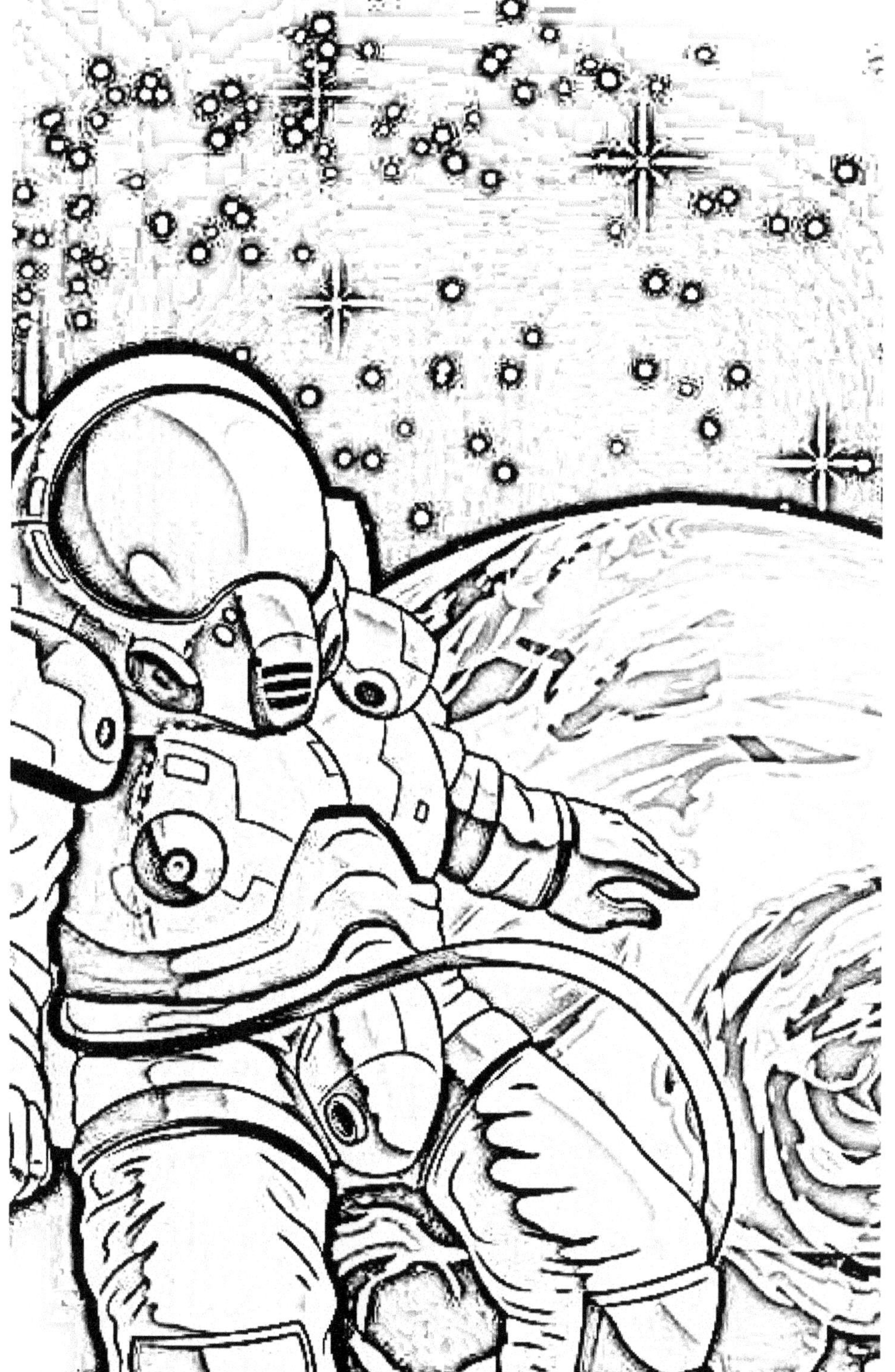

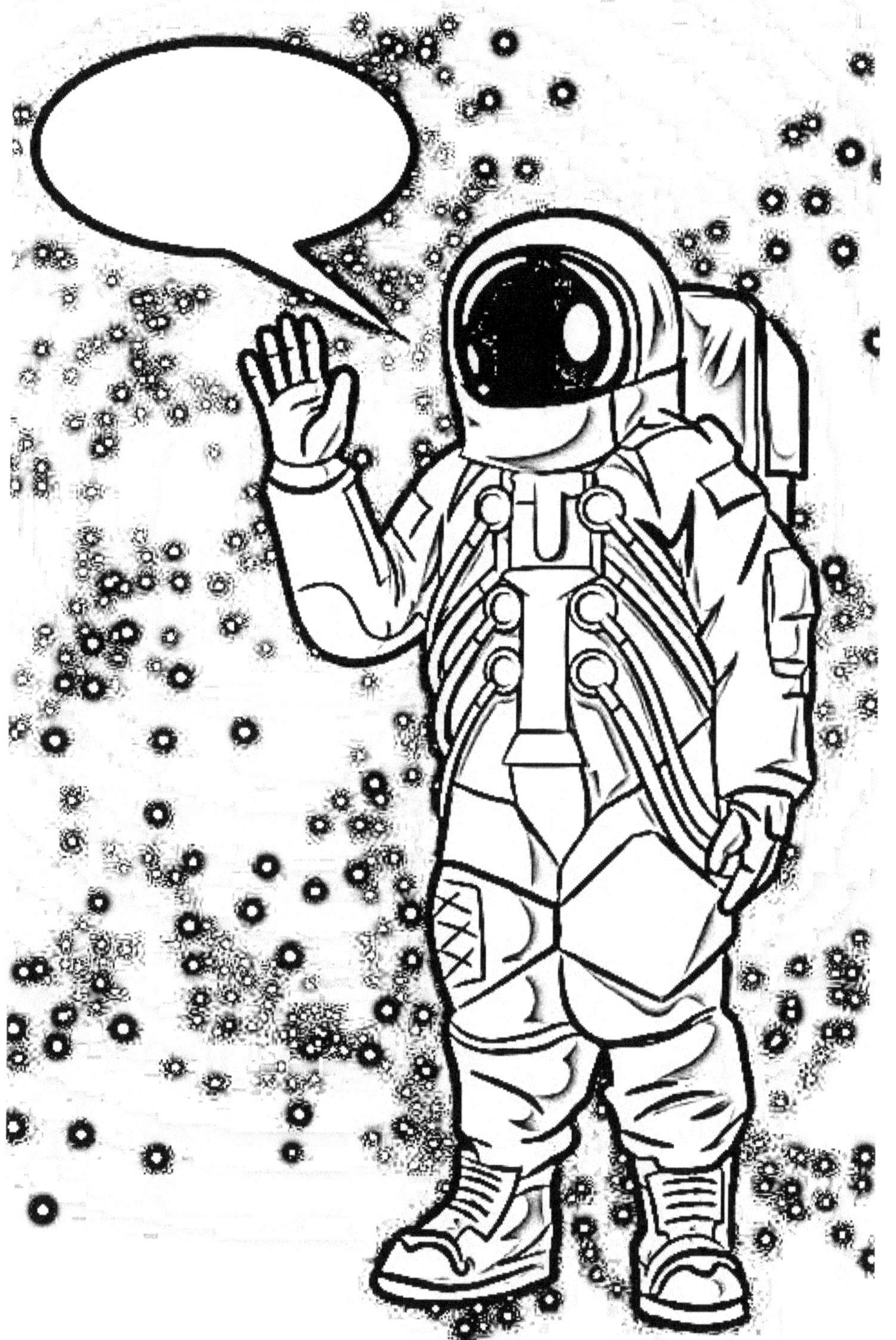

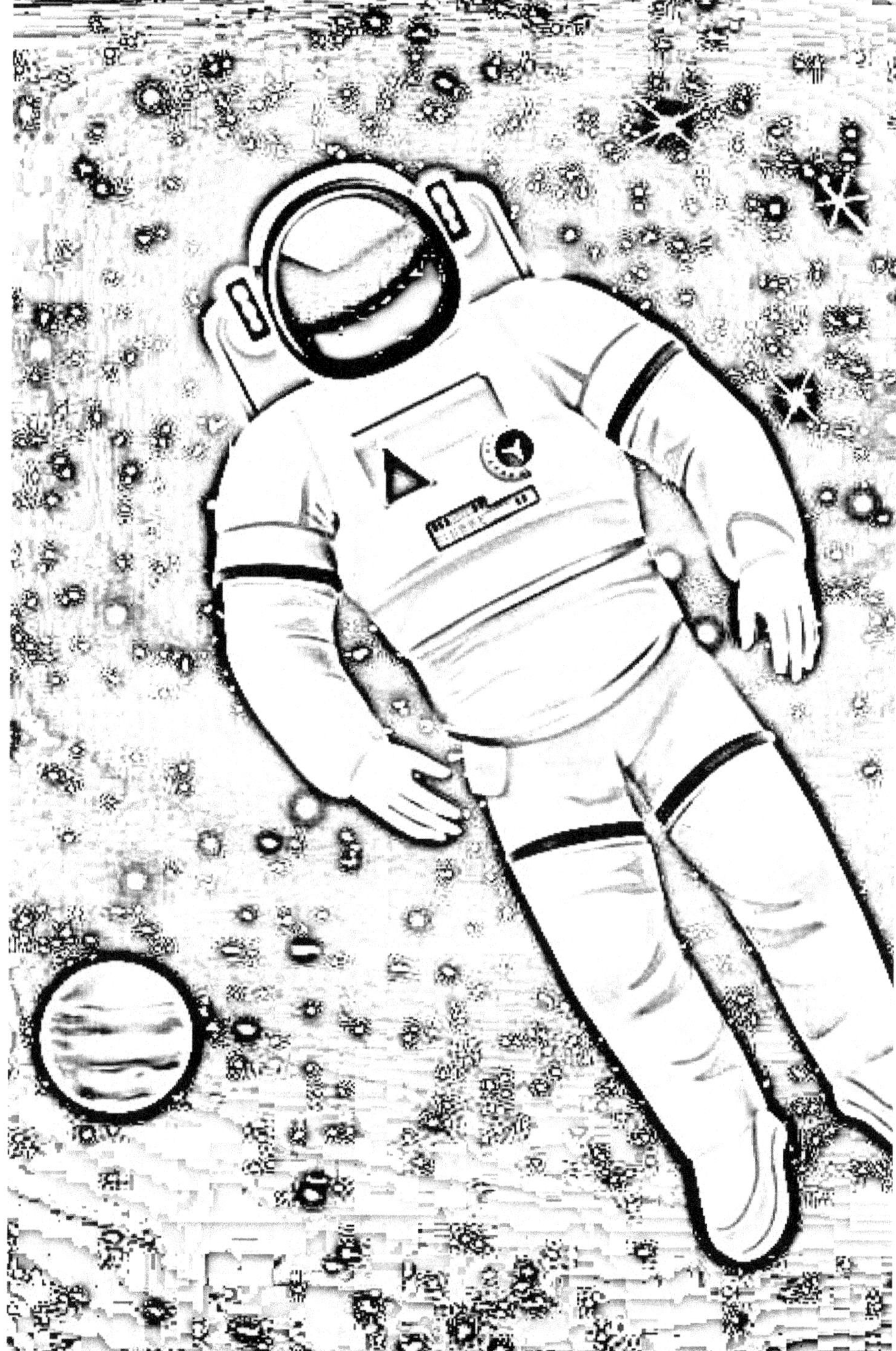

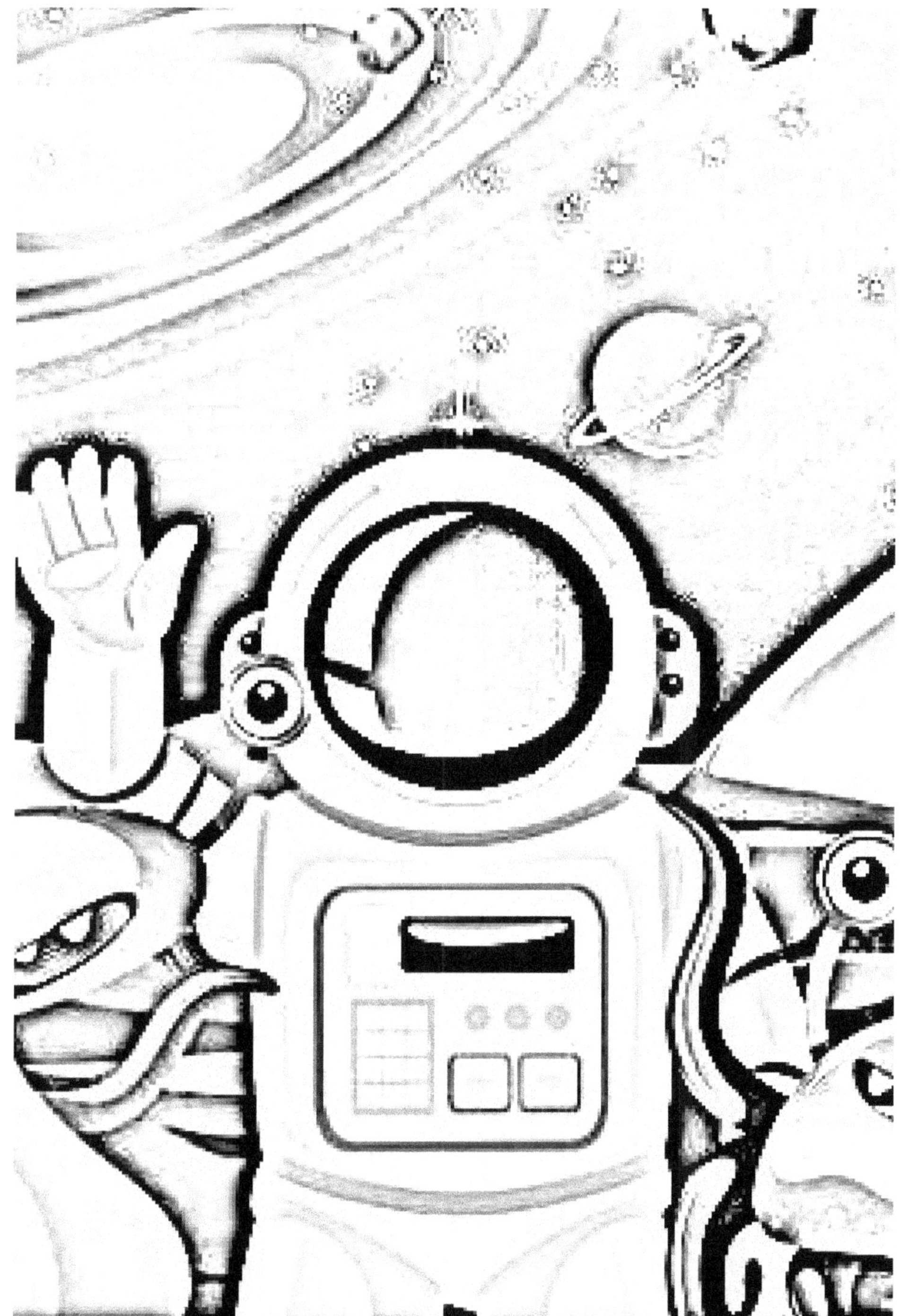

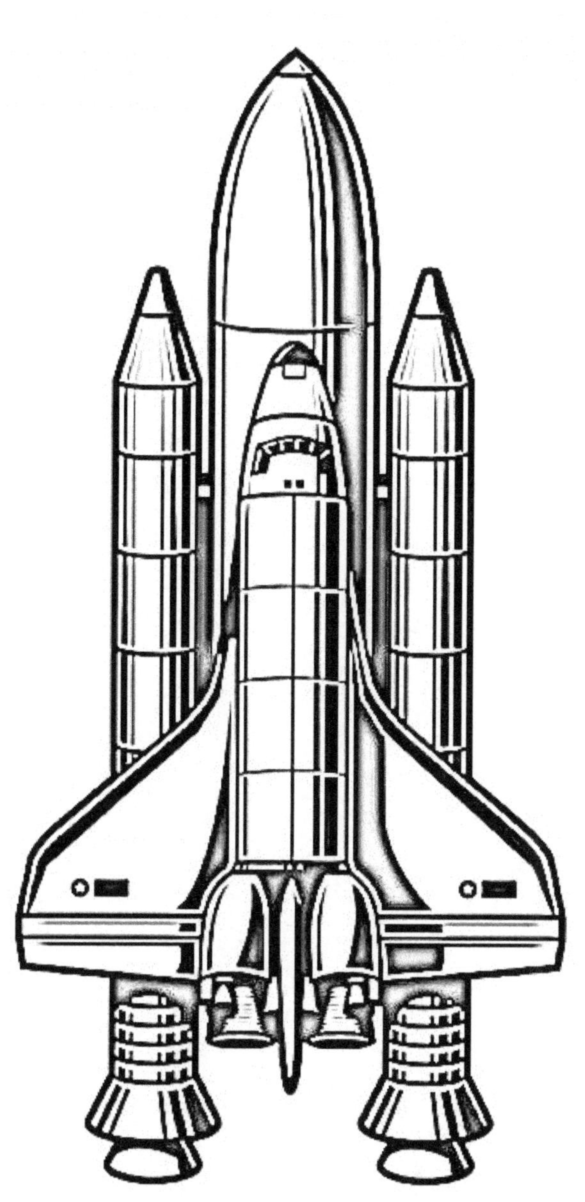

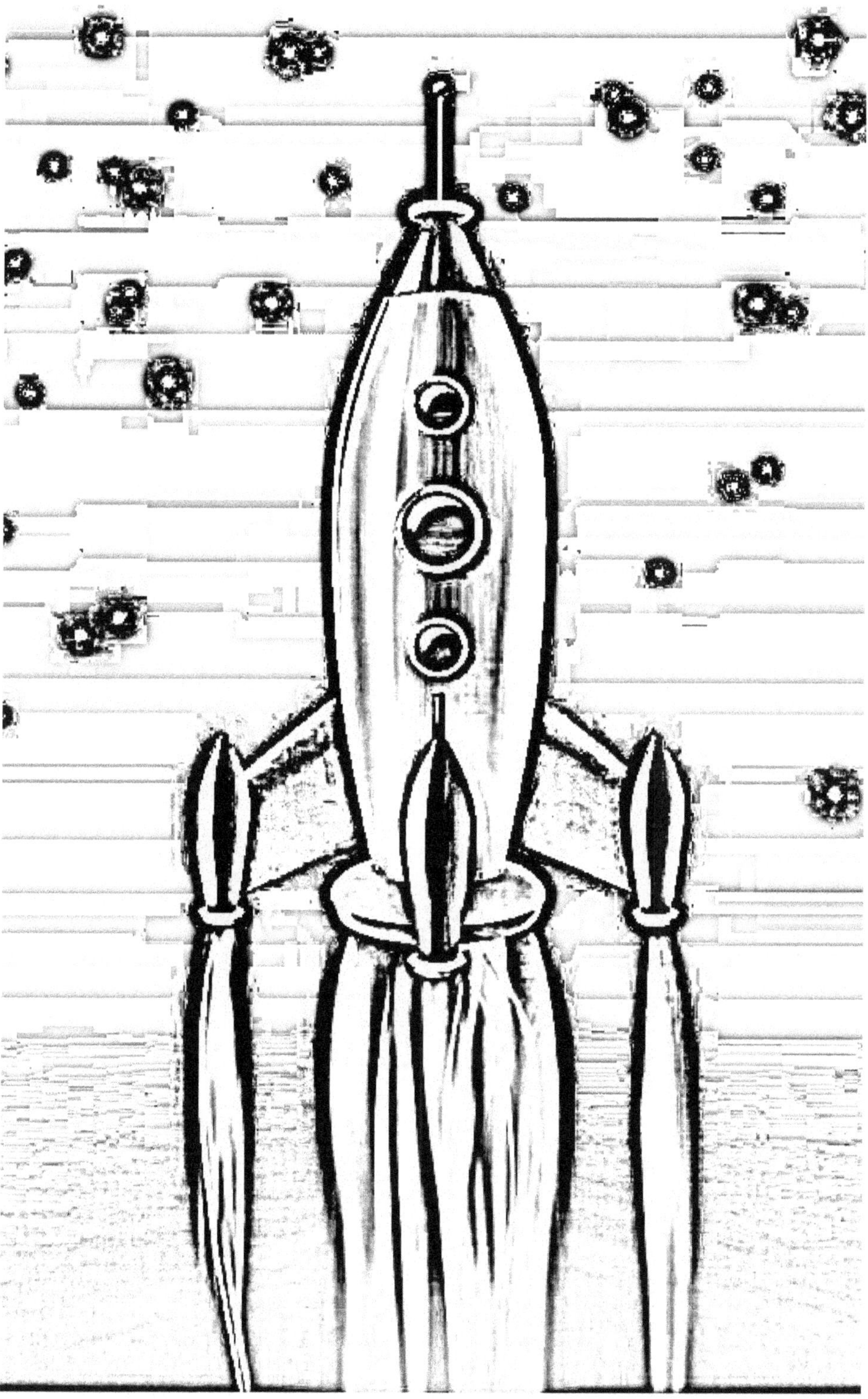

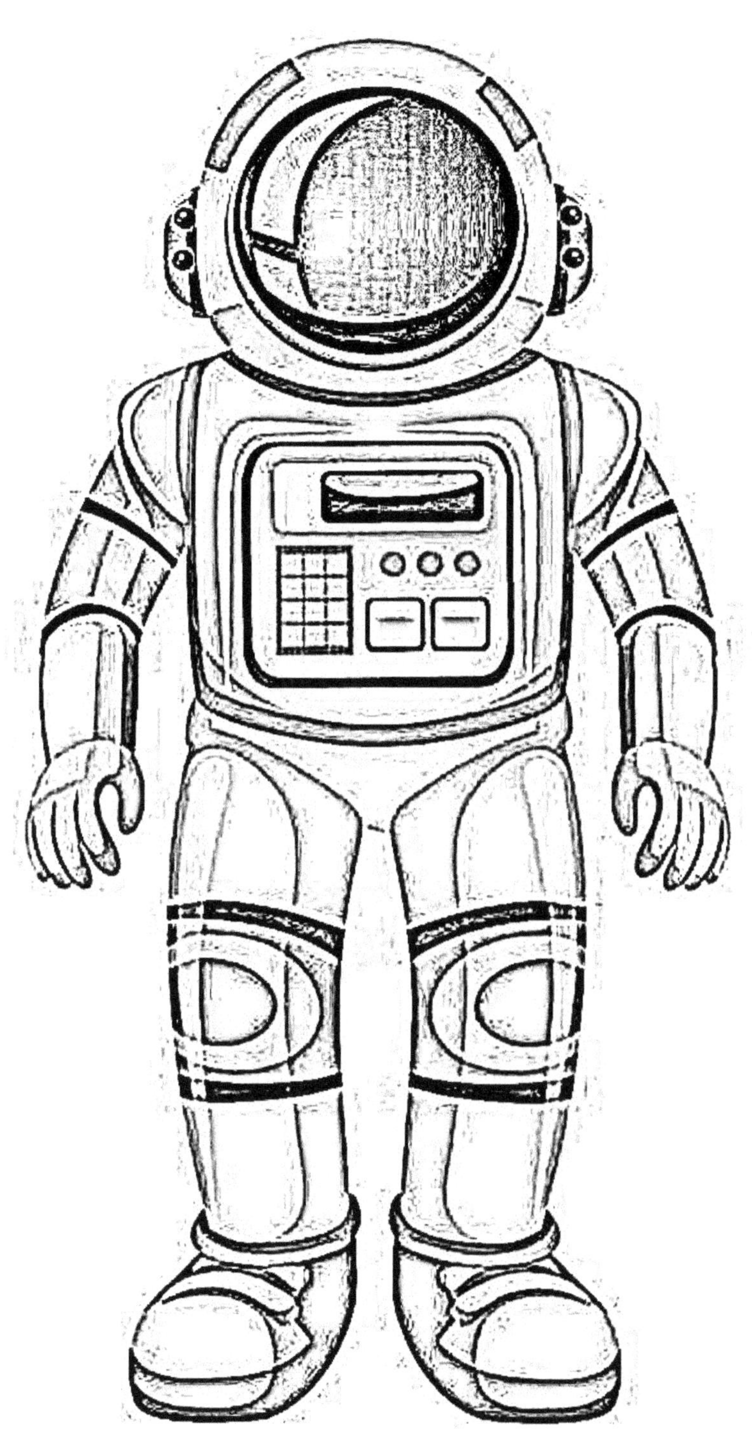

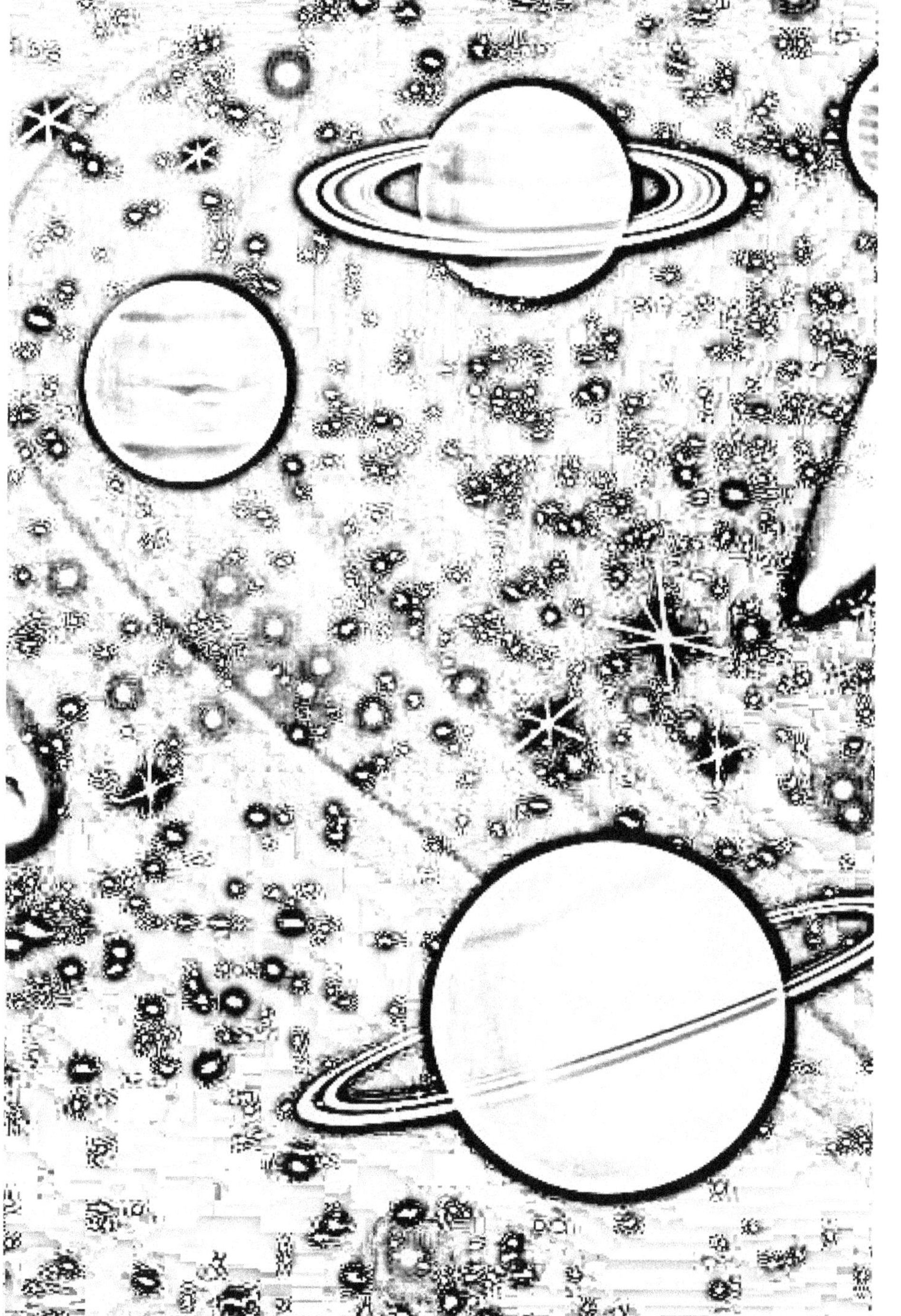